SCIENTIFIC POETRY

ECSTASY OF THOUGHTS

JYOTISHRAJ THOUDAM

To Richard Philips Feynman & Lord Krishna

Contents

Preface

India is rich in religious beleifs. To stand among the shoulders of giants, Newton also had to stand on the shoulders of some Christian readings to advance his calculus and theory. However, religion and science has been in a controverial relationship since Einstein, Hitchens etc. But to reconcile in the languge of science is my attempt, even though the stories of Vedas might or might not have been achieved, we were not there to witness it. But we can indeed recreate in different scenarios where the concepts described are achieved. I remember Prof. Arnapurna Rath's advice to look into concepts as they are the weaving thread of nature that runs like river among all fields. For that I thank her for my literary pursuit. Like those ancient ones, I also embeds scientific ideas inside these poetries, rather explicitly these time so that I can avoid Feynman saying "For the poets do not write to be understood" if he were alive. And the entries here are mostly didactic and sonnets. With this, I leave to the reader's knowledge in scientific fields of inquiry to my own "scientific orgasm" or "intellectual orgasm", enjoy!

Acknowledgements

My heart felt regards goes to Notion Press first, Prof. Arnapurna Rath, for she gave me the term "comparative literature" which encapsulated my entire career. Prof. Dilip Sundaram, for allowing me to be human and be free of choice and still allowing me to pursue my other endevours.

1. Electrical Brain

A changing electric field

Produces a magnetic field

And vice versa

So what do you want dear friends & foe

I need results

"Publishable, applicable results"

Now that's from the movie "A beautiful mind"

So what the fuck is yours dude? Are you a bore or what?

See, things that are moving keeps on moving

We live on the surface

Surface covered with non-intersecting lines of magnetic field

From mother Earth

As we walk

Our electric brains are happy, why?

They be charged from the changing magnetic field

Where? between the neuronal firings

Perhaps that's why ancient holy ones

Loved long walks

And even scientists like Perelman & Andrew Wiles

Perhaps the magnetic field be like

Hey! Let's help this hoooman's brain juice up

And then forget about it alllllllllllll :)

2. Quantum Mechanical Brain

Hey! Why don't we leave Maxwell for a while

And make friends with Schrodinger, Heisenberg & Dirac

But to breakup with Maxwell

I need Casimir's help

Why?

Hell! He's going to juice up our brain

Our Quantum Mechanical Brains

How?

Well!

Most of us are 99% vacuums

Or in the truest sense VOIDS

But Heisenberg disallows energy voids

How? By his pet Heisenberg's uncertainty principle

Saala, Energy time uncertainty principle

His pet is stronger than Aldous Huxley

Aldus Huxley, Darwin's bulldog for his Evolution theory, lol.

Perhaps Ramdas, Tulsi Das & Valmiki

Has hidden Fibonacci Ganeets in their writings

As I do in mine

Perhaps this is what we do

"We are what we repeatedly do"

Casimir effect will juice up the necessary

The necessary positron-electron pairs

Through one bark of Heisenberg's Dog of Energy-Time Uncertainty

3. Shit

I ate Neel gai (a deer type) shit this morning
I ate the dust near the Sabarmathi
Tasted the tips of the thorn
Gathered myself with the necessary natural pens
The sticks
All the above in the literal sense
For I am a budding scientist
Forever serving mankind
I cannot know everything
For that I need each of your help
Let's rise and take it from the unknown sources
And if we fall
Lets be fallen heroes
And the unnamed legends
For that I need your help
As dumbledore would say
"Help must be given to those who ask for it"
Or later he modified
"Help will be given to those who deserve it"
Or in my case
"Help must given to those who asks for it consciously knowing what
is true, reality and the unknown"

4. Expanding Mental Space

I once read

In an old Quantamagazine article

Written by Nobel Laureate Frank Wilczek

Titled "How Feynman Diagrams Almost Saved Space"

How Feynman Diagrams Almost Saved Space | Quanta Magazine

I linked the Quantamagazine article

I read that day, Feynman growled at Wilczek

To work on something other than "imaginary" Anyons (some weird particle name, which later was real & no longer imaginary)

I knew after that

How much love he had for particle nature of matter

He wanted to leave the wave nature of matter

He was sad, he couldn't fulfill Newton's exactness views of nature

Yet he was still unbowed

He continued to give the lecture in New Zealand

Praising the Dual nature of matter

Or to Mathematically put it, "The Probabilistic Point of View of Matter [PPVM]"

He praised Ernest Rutherford, and told the New Zealanders

Not to be sad of his presence

That they should be proud of Rutherford

He truly as "Physics with a human face"

And my "O captain my captain" along with Captain Algren ofcourse

He died, defending his enemy PPVM

For that he is a true scientist,

And without evidence it is not science

And with Heisenberg's bulldog, even Feynman cannot save space ;)

• 5 •

5. Bedtime Tales

My mother's eyes are like my own;

Shared in silence, in mum, over her selfless act of nine,

Utter less, this pain, she dropped on the first cry alone,

Her hold unmoved by the wind, nor the lifeless act of time.

Rigid is her manners and her caressing stance,

Bore this child's menacing and loving hells.

Resolute is her views brought by changing place and shifting face,

Seasoned is her advice, to lay placid among the mutation ails.

Led by the silencing echoes, and the bleeding paves,

Pushed this naive child into this changing waves of the world.

Now that the cradle is far and moving among vanishing days,

Only left to reminisce and lament on the longing yearns;

Of her calm, sweet and ushering voice moving this far, among life's noisy tales.

Pleasant like an unmade bed, warm like the evening Sun, do I still crave to hear her bedtime tales.

6. Kitchen Vibes

Kitchen Vibes

Music in my ears

Dipped spoon in my ugly coffee mug

Unopened coca cola that I bought last night

Perhaps people miss this things

The smell of the lpg from the left

Shining plates playing with my eyes

The uneven arrangements of my mother's utensils

I claim the following

Whoever has kept their crockeries is definitely not from India

The potatoes, trying to get inside the bucket

The tomatoes, pushing away the onions aside

I gather my peace and try to walk back to my room

But I wait until I finish this poem, bye

7. Laziness

Maybe laziness is not an illness
Perhaps its a gag reflex
Like that of the first cry of a child
A first cry of the child is not a voice
Rathe a sound released by the mere flow of air
However, is tastes sweet to the mother
Maybe it is nature's way of telling
That everything happens inside coexistence
Be it coexistence of liquids, solids and gas
Be it coexistence of four forces of nature
Or this might be the key to understanding "the naure of Nature"
Like Feynman put it in his Shykes documentary haha
He speaks jolly
He speaks merely
Without any care
Without any regret for completing any dare
The only cure to my laziness

8. Debt

Maybe life is full of cruelty

Where truth is crushed and beauty torn to shreds by the mighty shitteries of life

But at last I shall stand strong to mend my wounds

I am trying hard not to lick

But I am not like the many

I will not be trapped in the chasm of the many sins of men

For this heart beats for the innocent

For those beautiful beings and young hearts of this spherical surface

Well I do need money, be it prizes or change from the shop

But I will not hesitate to donate to the needs

And the ones who I see potentials

For which I have demonstrated so far

Without the need for recognition!

I bow to my army of brutes once more!

9. Exhausted

Well, the mind is more hurt than the body
When it comes to slight of hands
Maybe Feynman knew it
I am forever his gopi, in the Indian sense of the word
I ask consciously of his courage
I ask consciously knowing his capacity
I ask being aware of his mental faculties
Mixed with the kindness of Buddha
Intertwined with the healing power of mantras
Even though I do not agree to the supernatural nature of those
However I need it know
Therefore I call upon all the unknowns
To help me heal these unsaid wounds
Of many shits
For I know what it takes
The stuffs of brutes, naturals and gifteds
It is not given easily, It is made
And I say to the reader, build it for yourself
I submit that I am indeed mentally exhausted
But unbowed
Yet hopeful
And yes I shall rise again, for I am destined to be a scientis

10. Emptiness

Sometimes my brain doesn't let me sleep at night
Like a memory chip, I don't feel it
It just happens at sight inside
Perhaps NZT-48 would really fuck me up to it
Or perhaps captain Algren is my real captain
And not me
But this is just movie shit man
Why do I bother to be fucked up by thee
I don't even believe in that old guy
Or is it even an entity
I hated that idea, tried to snatch from their leaders or buy
But ideas are not real, I realized late in its entirety
Fuck it, I lie naked and sleep
Fuck it, If I die better learn something from it and sleep

11. Expression

Everything I breathe, touch, see, hear & taste

Is an expression of me

Where there is a plethora of imagined waste

There is certainly beauty untouched, so let it be

Well, maybe quotations fail to grasp the essence of our own truths

We must although try to fit in

To join the collective force of brutes

But without loosing the self in the many sin

We shall fail to express

In connection to the outside objects

But that doesn't mean we have failed to achieve progress

We can always look into our own vices of rejects

And be content with our own beings

And be switched into the mode of our doings

12. Sight

An utterless sight, leads to bottomless plunder,
Visions of glory, makes a soulless seeker.
Tapestry of lies, revealed by the maker,
Stories of hypocrisy, weaved by the painter.
Akin to mummify, this form, laid bare for long,
Refrained from release, this beast of light.
Seasoned with pain, a salted breath tasted strong,
Holding on, anguished, with a patient fight.
Accustomed to dodge, this swift fury of might,
Rage Rage, but calm outside.
Fear of distance, by the many light,
Kept this posture, for the fickle side.
This untrained judgment paused for those difficult sight.
This seasoned vision for the capricious fight.

13. Appetite

Like an appetite for the mind
I do not indulge in it
For it is left to carve a beauty inside the natural line
Cannot expect everyone to tune with it
For the easy is preferred
And the hard ignored
For the complicated is left to be pondered
And the lines of sanity is hard-won without a record
I can say with utmost certainty
Without any stains or pigments
It is the appetite for the mind to be checked for eternity
It is the will of men to be tested for many resentments
Shoulds, Musts & Forced, my well-informed enemies
While the Needs, Wants & Could, my old remedies

14. Imperfections

I like imperfections & mistakes
They make me feel safe
For if it were perfect
And it fails
I would not know how I'd survive
Without knowing the reasons for its failure
For the reasons of the unknown
Is what makes men mad
I would become mad
If not for the imperfections caressed me deeply
As I lay here, in her bosom of carelessness
I could see the light
Through the cracks of these imperfections
Through the thorns of these bleeding mistakes

15. Night

The night is lovely deep and dark
Where mankind's bitterness vanish among the silence
This bites of loneliness and slipping peace
Maybe this is life, A way around the ways
Where mundanity exist inside happiness
Revealing itself only to the unaware and ignorant
To the unconscious and the closed eyes
Maybe there is hope, hope to live and vanish along
With the flow, that pierces into the deep black night.

16. Simple Pleasures

The smell of the gasoline numbing the senses,
Not thinking when to stop.
The slow boiling sound of cabbage and potato,
Brimming from the old earthen pot.
The slow wind tonguing your body,
Playing with your hair, while you try to gulp that fresh breath of air.
The sound of thunder growling quietly,
With the voice of the rain beating continuously singing a lullaby.
Maybe I am lost, lost in the boiling broth of earthly pleasures,
Where most mortals ignore for a soothing piece of meat,
Where there is plenty of delights yet to be explored.

17. Dedication

Hello!

Nikola Tesla (Denied Recognition!), Freeman J Dyson (The fucking Rebel without A Ph.D.), Arnold Sommerfeld (Most nominated but denied the Nobel), Thanu Padmanabhan (Didn't live to see the medal) Jean-Pierre Serre (Field's Medal 27) , Malala Yousafzai (Nobel 17), Lawrence Bragg (25 Nobel), Maharaj Kumari (MK) Binodini (where is she among the WORLD'S LIST?)

We are here, to make you the unnamed legends proud

You all above deserve to be prayed like GODS!

Unlike the Naturals! & the gifted lazies & the complacent geniuses

Let it be known from this day 7:55 AM 3/3/2022 (Takyel)

We, the army of brute forces is here to give you the recognition you deserve

The young brute forces who deserves to rise among the heap

To lead the hive like Jean-Pierre Serre, Mala Yousafzai & Lawrence Bragg

The likes of Abdus Salam, who was denied recognition by Wolfgang Pauli

His letter on symmetry (later some two guy won the Nobel with the same theory lol :(he deserved it earlier) was dismissed early by Wolfgang Pauli

That's why he advised the young researchers to distrust established authorities :)

Don't worry dear armies of brute forces

We stand strong, you all are remembered from us

As long as we unite with forces STRONGER THAN THE 4 FORCES OF NATURE

You all will be remembered for you did what you think was RIGHT!

FOR FEYNMAN SAID "DISREGARD! For established authority" AFTER HE READ "DOUBLE HELIX" by JAMES WATSON!

We BOW TO YOU AND WILL BE REMEMBERED FOREVER AND EVER!

✌??????????????♀??♂??♂????????♂?

(8:07 AM, 3/3/2022 Sagolband Takyel Kolom Leikai, Imphal West Manipur, India, 795001)

Epilogue

To create scientific orgasm, I continue to propose that ideas must be mixed and should not create a separation of boundaries. For boundaries create a feeling of strait jacket to the imagination. Although Richard Philips Feynman said, "Science is imagination in a tight strait jacket", I would improve upon this by the following propostion that "Science is imagination inside the brain, with memory space of the entire multiverse taught to us by the so called wink wink Godhead Krishna, and polishing & remaking realistic tools to achieve it in reality and not just in imagination and thoughts". With this I bow out and head off to other endeavours untill I come back here for more!